DISCARD

Cryobiology

Cherie Winner

LERNER PUBLICATIONS COMPANY
MINNEAPOLIS

To my dear friend Wendi Silvano

The author thanks Dr. Richard Lee of Miami University, Oxford, Ohio;
Dr. Carlisle Landel of The Jackson Laboratory, Bar Harbor, Maine; and
Dr. Brian Barnes, director of the Arctic Research Institute, University
of Alaska, Fairbanks, for sharing their knowledge of this cool field.

Lerner Publications Company
A division of Lerner Publishing Group
241 First Avenue North
Minneapolis, MN 55401

Website address: www.lernerbooks.com

Library of Congress Cataloging-in-Publication Data

Winner, Cherie.
    Cryobiology / by Cherie Winner.
        p.   cm. — (Cool science)
    Includes bibliographical references and index.
    ISBN-13: 978–0-8225–2907–1 (lib. bdg. : alk. paper)
    ISBN-10: 0–8225–2907–6 (lib. bdg. : alk. paper)
    1. Cryobiology—Juvenile literature. I. Title. II. Series.
    QH324.9.C7W56 2006
    571.4'6451—dc22   2005006158

Manufactured in the United States of America
1 2 3 4 5 6 – BP – 11 10 09 08 07 06

Table of Contents

Introduction

One of the coolest sciences around is cryobiology. *Cryo* means "cold," and cryobiology is the study of life at low temperatures. Cryobiology is more than learning how animals stay warm. In the hot new field of suspended animation, scientists are looking for ways to use cold to put life on hold. Suspended animation is like hitting the Pause button on a DVD player.

It's putting someone in the deep-freeze and thawing that person out—alive—many years or even hundreds of years later.

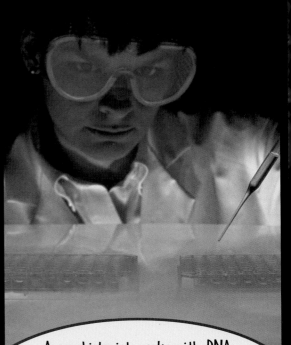

A cryobiologist works with DNA (deoxyribonucleic acid), the molecule in the nucleus of most cells that holds genetic information. She is preparing the DNA for cryogenic storage (deep-freezing).

Actress Sigourney Weaver sleeps in a cryochamber in the movie *Alien*. The chamber keeps her alive while she hibernates during a flight into deep space. One day, cryobiologists may find a way to make this Hollywood fiction a reality.

It's making an astronaut sleep through a 20-year flight to another planet.

It's keeping hearts and kidneys alive until they can be matched with patients who need a transplant, no matter how long it takes.

We can't do any of these things yet, but we're getting closer all the time. Our best clues for how to suspend animation come from animals you might see in your own backyard, such as chickadees and ladybugs. For us, suspended animation will be a great achievement. For them, it's a normal part of life.

chapter 1

Hot Possibilities

Many wild animals can put their lives on hold. They have to or they would die. For them, suspended animation is a way to get through hard times, such as nasty weather or lack of food and water.

When wild animals face harsh conditions, they have three choices. They can stay active and struggle through the rough times. They can go someplace where life is easier. Or they can put their lives on hold and wait until things get better.

This queen hornet is hibernating in a piece of rotting wood for the winter. In spring, warmer temperatures will wake her from her sleep.

IT'S A FACT!

Cold isn't the only reason animals go into suspended animation. Some species shut down because of intense heat or because their environment dries up.

Look at how animals deal with winter. Many birds and mammals stay active no matter how cold it gets. They keep warm by using food as fuel. It also helps to be covered with blubber, fur, feathers, or long underwear!

Some animals, including monarch butterflies and many birds, migrate to warmer places where food is easy to find. They return in spring, when the weather warms up and food is abundant.

Two male polar bears play-fight during the winter. Male polar bears are active all year. But most females spend the winter nestled in dens with their young. When spring arrives, females and cubs leave their dens.

Other animals stay in their cold environment, but they go dormant. That means they're not active. They're in suspended animation. They stop eating and drinking. They don't go to the bathroom. They may not move at all. Sometimes they don't even breathe. Their metabolism, the total of all the chemical processes in their body, slows down or stops.

Dormant animals often look dead. But when the right conditions come along, they revive, or "come to life" again. How do they put their lives on hold for weeks or months? Could we do the same?

When the Going Gets Tough, the Tough Go to Sleep

Different animals go dormant in winter in different ways. Some insects and frogs freeze. Their body fluids turn to ice. If that happened to you or me, the ice crystals would tear our cells and tissues to shreds. When we thawed out, we'd be mush. (And, of course, we'd be dead.) But these animals survive just fine.

Ice crystals form on a bat (*left*). This is normal. In spring, the bat will warm up unharmed. But exposure to such cold for humans (*right*) can be deadly.

Nothing to Eat

Many animals that go dormant in winter don't do it to avoid the cold. They do it because their food disappears. Animals that eat seeds usually stay active all winter. Animals that eat other animals, insects, leaves, or fruit must migrate or go dormant.

A timber rattlesnake (below) is hibernating because its prey, such as this dormouse (right), are hibernating or have migrated. Rattlesnakes, which can hibernate up to six months, also go dormant to survive cold temperatures.

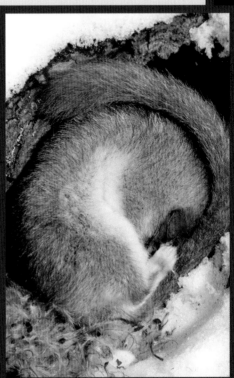

Some insects don't freeze, even when their environment gets very cold. Instead, they supercool. Their bodies make antifreeze chemicals that stop ice from forming inside them.

Some mammals and birds hibernate for weeks or months. Others go dormant for shorter times. They let their body temperature drop at night, when the outside temperature drops. This is called controlled hypothermia (a body temperature lower than usual), because the animals can turn it around and get warmer again. In the morning, when the sun warms things up, the animals raise their temperature back to normal.

Dead or Alive

Here's a question for you. If an animal doesn't move, has no heartbeat, and is not using any oxygen, is it dead? Or is it dormant? Sometimes the only way for even a scientist to tell is to put the animal in better conditions and wait to see if it "wakes up."

A keeper holds a long-eared jerboa shortly after its hibernation at a German zoo. Among other things, the zoo is observing the rodent's hibernation patterns.

What about Us?

People don't go dormant at all. Sleeping in on a Saturday morning doesn't count!

No matter how cold our environment is, our metabolism tries to keep our temperature at around 98.6°F (37°C). If it fails, we're in trouble. We can't survive being frozen, we don't super-cool, and we

Humans can't hibernate. We must use warm clothing, shelter, and heat to adapt to cold temperatures (*top*). But human infants have a kind of fat found in hibernating animals (*bottom*).

don't hibernate. If we go into hypothermia, we need help to warm up again or we will die.

We don't naturally go dormant. But we can learn from the animals that do. In this book, we'll see how hints from wildlife are helping scientists solve the puzzle of suspended animation.

Let's start by taking a closer look at those frosty frogs.

Life on Ice

On an October evening, a wood frog hops into a pile of leaves. It has spent the last few nights in similar places, protected from frost. This time the frog nestles deep into the leaves. It's going to stay put for the winter.

Over the next two days, the temperature falls and stays below freezing. The frog gets colder and colder. Finally, a crystal of ice forms on its leg. Within minutes, the ice pierces its skin. The ice spreads quickly along its muscles. As the water in its body turns to ice, its internal organs shrink and its eyes get cloudy. About 24 hours after it started to freeze, its heart stops beating.

But the frog is not dead. When warm weather returns, it will "wake up." Even before the frog's legs can move, its heart will start beating again. A few hours later, the frog will emerge from its pile of leaves and hop away to hunt for breakfast.

This wood frog has found a safe place to hibernate. It has the awesome ability to freeze, thaw, and awaken unharmed.

Even though wood frogs freeze naturally, no one has been able to freeze a wood frog heart, then thaw it out and have it work right.

What about People?

People who freeze don't wake up. They die. Our body parts can't stand being frozen at all.

That's why frostbite is so dangerous. Imagine you're stranded outside in winter without gloves. The colder your fingers get, the less blood they receive. Soon the cells in your fingers run short of oxygen. They can live without oxygen for just a few minutes. After that, they die or get badly damaged.

When your fingers get cold enough, they start to freeze. Ice crystals form first in the spaces between cells. Later, ice crystals form inside the cells as well. Their sharp edges rip holes in cells, blood vessels, and other delicate structures.

When you warm up, you face more problems. Blood returns to your fingers, but it can't circulate (flow) because all the tiny blood vessels have been shredded by ice. The blood flows in but can't flow back out. The extra blood makes your fingers swell. Any cells that survived being frozen get squashed. They die, and your fingers must be amputated (surgically removed).

A wood frog comes out of hibernation, perfectly healthy with skin intact (*above left*). Freezing, such as what happens during wood frog hibernation, damages human tissues (*above*).

Freezable Frogs

Why can wood frogs freeze their whole bodies, but we can't even freeze our fingers? What do they do that we don't?

Wood frogs flood their bodies with cryoprotectants. These chemicals protect the frogs from chill damage and sharp ice crystals.

Many cryoprotectants are common chemicals, such as glucose, a natural sugar in blood. When a frog freezes, its blood carries much more glucose than usual. If you had that much glucose in your blood, you would have the disease called diabetes. If your blood glucose stayed that high for more than a few minutes, you would die. Our cells need small amounts of glucose, but large amounts are toxic (poisonous). But the extra blood sugar is not toxic for the frog.

Bits and Pieces

So far, no one has been able to freeze and revive a human heart or lung, let alone a whole person. Freezing an organ is easier than freezing a body, but it's still very hard. One of the biggest problems is finding a safe cryoprotectant. Many cryoprotectants only work when they're present in huge amounts. Then, similar to what happens with glucose, they become toxic to human cells.

Chill Damage

Low temperatures hurt our cells even if the cells have enough oxygen and don't freeze. The skinlike membrane that surrounds each cell gets leaky. Then it can't keep the cell's stuff in and other substances out. This is called chill damage.

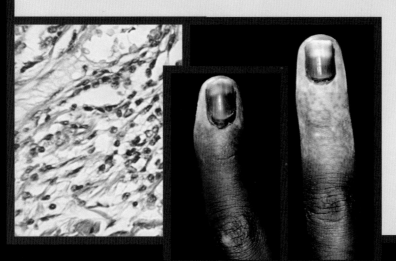

Chill damaged fingers and a microscopic view of chill damage at the level of the cell (left)

Organs are also tricky to freeze because they contain many different types of cells. Some cells need one kind of cryoprotectant while others do better with a different kind. How do we get the right amount of the right cryoprotectant to all the cells? The size of the organ is a problem too. It's almost impossible to cool all parts of the organ evenly. If outer parts get frosty while inner parts are still warm, the fragile organ can rip open.

We have better luck with smaller pieces of tissue. Corneas (the outer covering of the eyeball), heart valves, and embryos (unborn, developing animals) can be cryopreserved. They are frozen and stored in liquid nitrogen at −320.8°F (−196°C).

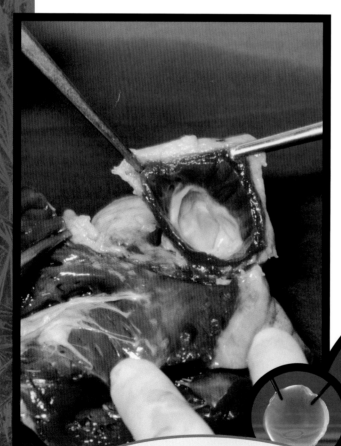

We're also good at freezing blood cells and sperm. These are cells that aren't connected to other cells in a tissue or organ. Each cell works on its own. Sperm were the first cells we learned to freeze and revive, back in 1949.

A surgeon removes the aorta (a blood vessel) from a human heart (*top*). Another doctor handles a human cornea (*right*). The aorta and the cornea will be cryostored until patients need them.

Freezing Organs

One reason organs are hard to freeze is that they can't afford to lose many cells. Sperm are different. Many sperm die when they're frozen and thawed. But if a sample starts with 1 million sperm and 95 percent of them die, the sample will still contain 10,000 living sperm. That's more than enough to fertilize a few eggs. But for a heart to work right, almost all of its cells have to be healthy. If a heart loses more than a few cells, it can't be used for a transplant.

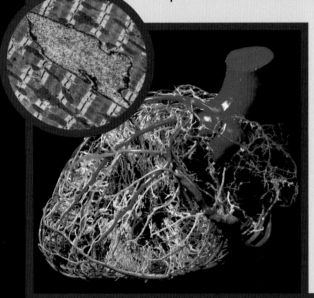

The heart (left) is a complex organ, with many vessels and chambers made up of cells (top). Currently, this complexity has prevented cryobiologists from freezing and thawing a healthy heart.

Sperm are frozen like this: A sample is placed in cryoprotectant fluid inside a hollow straw. It's cooled to between −31 and −112°F (−35 to −80°C). Then the straw is stored in liquid nitrogen in a metal vat called a dewar. Dewars are like giant thermos bottles. They keep samples cold without using electricity. When the sperm are needed, they are thawed at room temperature and the cryoprotectant is washed out.

The Mouse Bank

Some of the coolest freezing research is happening at The Jackson Laboratory in Bar Harbor, Maine. Scientists there freeze mouse sperm and embryos. These can be used to produce living mice.

Cells, Genes, and DNA

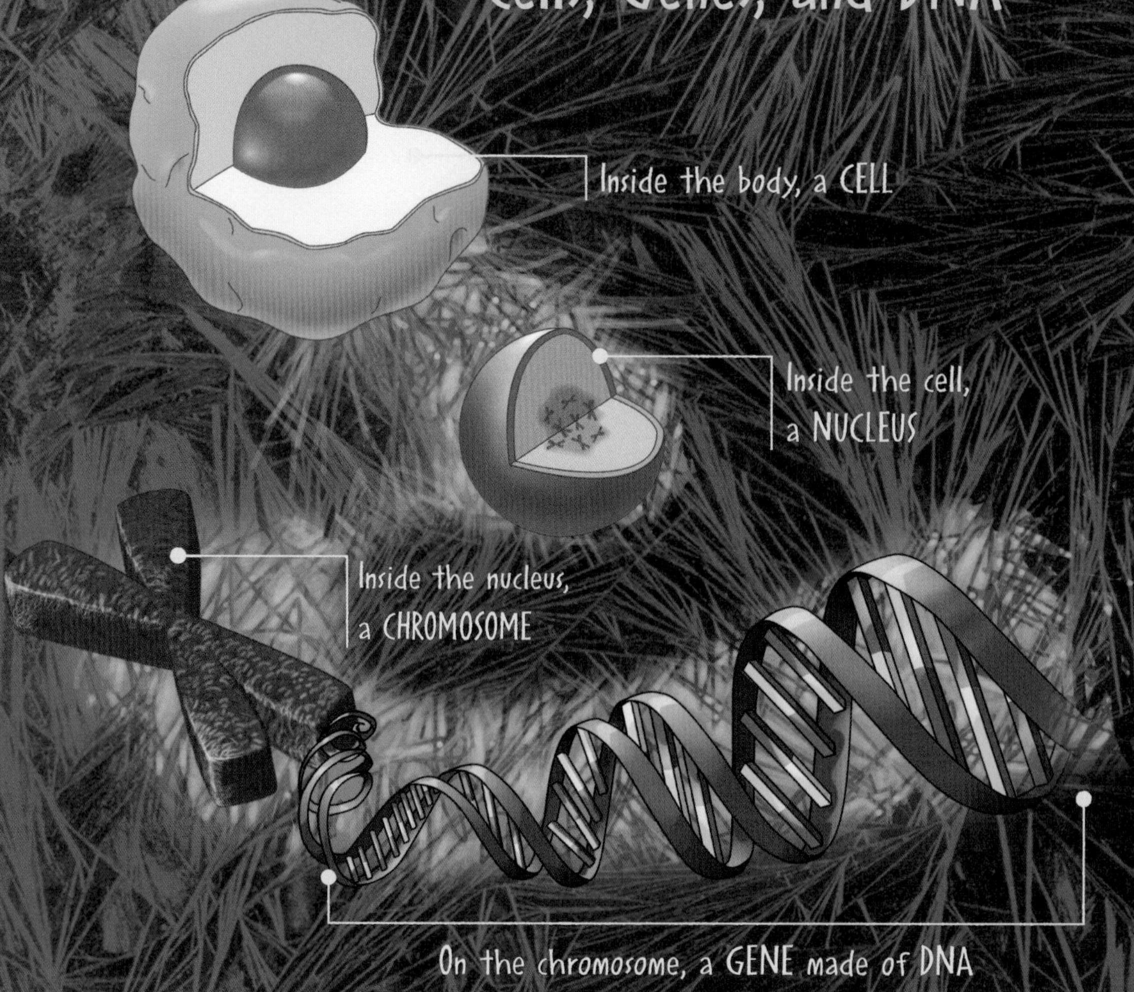

Inside the body, a CELL

Inside the cell, a NUCLEUS

Inside the nucleus, a CHROMOSOME

On the chromosome, a GENE made of DNA

These aren't just any old mice. They have been bred or engineered (genetically changed by humans) to have genes with mistakes in them. Genes are the chemical "blueprints" for all the structures in the body. Some of the mice carry genes that give them the disease sickle-cell anemia. Others have cystic fibrosis (CF) or a brain disease or some kind of cancer.

Scientists all over the world use mice from The Jackson Lab to try to figure out how a disease works and how we might treat it. For instance, scientists might test a new drug for CF on mice that have the disease before trying it out on human patients.

Freezing mouse sperm and embryos until needed by scientists costs a lot less money and takes up a lot less space than keeping mice in cages. It's also safer. In 1947, before the Jackson Lab started freezing its mouse sperm and embryos, the whole lab burned down. All their mice were killed. After the fire, the lab began keeping backup samples at other places.

What about Whole Bodies?

You might have heard the word *cryonics.* That's the word for freezing whole people, in hopes that they can be revived someday. About 100 human bodies are already in cryonic storage in the United States. Most of the frozen people died of diseases, such as cancer or heart failure. They had themselves frozen right after they died. (It's illegal to freeze someone who is still alive.) They hoped that in the future, doctors will be able to bring them back to life and cure their disease.

A man displays his medical alert bracelet (*below*). The alert tells medical staff that he wishes to be cryogenically stored (with the hope of being revived in the future). The outlook for his success isn't good. Currently, cryogenic storage destroys the cells of large human and animal organs. The tissue at left is from a cryogenically stored and thawed dog brain. The open areas show tissue damaged by ice crystals that formed during freezing.

By now you have figured out that people who have been frozen probably won't wake up again. They will need much more than the cure for the disease that killed them.

CALL NOW FOR INSTRUCTIONS
PUSH 50,000 U HEPARIN IV &
DO CPR WHILE COOLING WITH
ICE TO 10C. KEEP PH 7.5
NO AUTOPSY OR EMBALMING

Ready for a Total Body Washout?

Are you sure you want to go into the deep freeze? Here's how it works: As soon as you die, cryonics technicians cover you in ice to start cooling. They open blood vessels to drain out your blood and replace it with ice-cold cryoprotectant fluid. They drill a hole in your skull so that they can watch the brain to make sure it is absorbing the cryoprotectants and to take temperature measurements. After the surgery, you're placed in liquid nitrogen at −385°F (−196°C). That's where you'd stay until doctors could cure what killed you and restore you to good health.

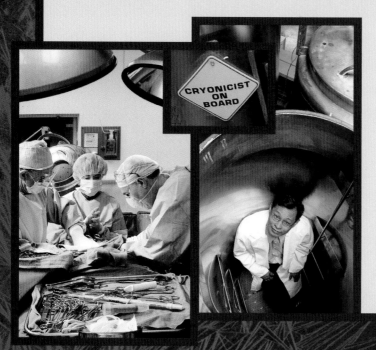

Doctors at Alcor, a cryonics company, prepare a human body for cryostorage (far left). The president of Trans Time, Inc., stands in a cryochamber (left).

No matter how carefully they were frozen, almost every part of them will be damaged by ice or cryoprotectants. Even if they do wake up, they will need repairs to almost every body part.

What's next? Let's leave the freezer and check out an animal that gets through the coldest winters without freezing.

IT'S A FACT!

The Jackson Laboratory uses sugars as cryo-protectants. It also uses a "magic ingredient": skim milk. Scientists don't know how it works, but skim milk helps the cells survive being frozen.

chapter 3

Super Cool

In a tree above the wood frog's leafy bed, a ladybug clings to a twig. All through the winter, it will be exposed to the icy air. It will get colder than the wood frog, yet the small beetle will not freeze.

Instead, the ladybug will super-cool. Water usually freezes at 32°F (0°C). But water in the ladybug will remain liquid at temperatures down to 3.2°F (−16°C).

While the ladybug is super-cooled, its life is totally suspended. Its heart doesn't beat, and the ladybug doesn't breathe. It's a lot like being frozen, except the beetle does not turn to ice.

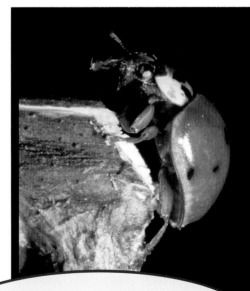

This ladybug is clinging to a twig in preparation for hibernation. The beetle will be exposed to freezing temperatures, but it will not freeze.

There are two secrets to supercooling. (Well, at least two. These are the ones we know about so far.) First, animals that supercool make antifreeze chemicals, or ice blockers. Some antifreezes prevent ice from forming at all. Others let tiny ice crystals form but then bind to them so they can't grow any bigger. All ice blockers work best at certain temperatures. If the temperature goes too low, they won't work. Big ice crystals will form, and the animal will die.

The second secret is more complicated. The animal has to stay away from anything that might act like a "seed" for ice to form. A seed, or nucleator, is any tiny speck that water can cling to. Dust, bits of food, and bacteria can all be nucleators. So can ice that has already frozen.

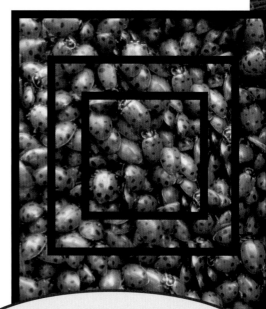

Hundreds of ladybugs in hibernation. These beetles carry their own antifreeze, allowing them to survive supercooling.

Nucleators help the water molecules line up in just the right way to form a crystal of ice. Without a nucleator, the water molecules won't bind to each other in a way that forms a crystal.

Staying away from nucleators is hard because they are all around in the environment and inside the body. Going into winter, a ladybug stops eating so its stomach will be completely empty. It even poops out the microbes that normally live in its gut. Its tough, waxy skin protects the beetle from ice in the environment. If a nucleator stays inside it or if one pokes through its skin, it will freeze in a flash and die.

Supercool Squirrels

Most animals that supercool are very small, because the bigger the animal, the harder it is to get rid of all the nucleators inside it. Arctic ground squirrels are

the largest animals known to supercool. They are also the only kind of mammal known to supercool. The Arctic ground squirrel drops its body temperature to about 27°F (−2.9°C) when it hibernates. We don't know how the squirrel keeps from freezing or how it survives being that cold.

An Arctic ground squirrel hibernates in its burrow. Arctic ground squirrels hibernate as long as seven months.

Lollipop, Lollipop

Scientists are beginning to explore how to supercool organs so they can be saved for transplantation later. But scientists are also doing something even the ladybug doesn't do.

If you add cryoprotectants and ice blockers to water and then cool the mixture to −202 to −238°F (−130 to −150°C), it does something really weird. It becomes a glass! (Window glass, made of silicon dioxide, is another kind of glass.) The water becomes hard, but it doesn't contain ice crystals that could damage sensitive cells. This is kind of like making a lollipop out of sugar water.

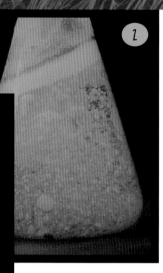

The process of making a glass is called vitrification, from the Latin word for "glass." It's one of the most promising ways of making organs go dormant. Doctors are working on finding the right mix of cryoprotectants and ice blockers to use with different kinds of organs.

Clockwise from left to right: Flasks showing frozen and vitrified solutions. Flask 1 has no ice blocker. The solution is frozen. The addition of some ice blocker to the solution reduces ice crystals (flask 2). Flask 3 contains the most ice blocker and the least amount of ice.

IT'S A FACT!

In liquid water, there is room for water molecules to move around and bump into each other. In an ice crystal, the molecules bind to each other and are "frozen" into position. In vitrified water, the water molecules are held in place by cryoprotectants or ice blockers.

Warming up a vitrified organ is even more dangerous than cooling it down. Ice is more likely to form during the warm-up. So an organ might be vitrified safely, but as scientists start to warm it up, ice might form inside it and tear it to pieces.

The warm-up has to be done very fast—at least 36°F, or 20°C, per second—to prevent ice crystals from forming. So far, the best way to warm vitrified organs is with microwaves. These waves of energy are like the ones used in a microwave oven but at a much lower power.

Unfortunately, just like the waves in your microwave oven at home, these low-power microwaves have a problem. They sometimes create hot spots and cool spots. Then the organ may split open, like when a burrito "pops" in your microwave.

Storing organs is great, but what about making whole people go dormant? We're still a long way from being able to safely freeze or vitrify a living person. What if we skip the ultracold temperatures and get just cool enough to slow down and go dormant?

Glass Organs?

In 1997 scientists discovered how to vitrify human blood vessels and other small tissues. Scientists are close to being able to vitrify rabbit kidneys. Soon they may be ready to turn human organs into glass.

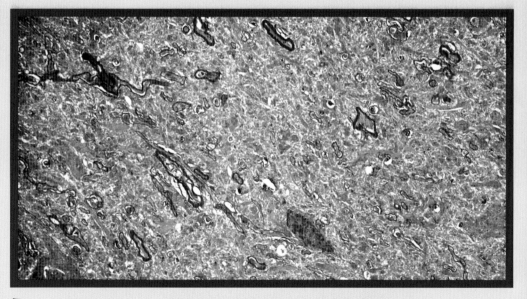

This microscopic view of rabbit brain tissue following vitrification and warming shows little cell damage. Scientists used an ice-blocking solution that seems promising for use in human tissue vitrification.

chapter 4

Chilling Out

While the wood frog turns to ice and the ladybug supercools, chickadees go dormant part-time. During the day, the flock forages for seeds. In the evening, the flock splits up. Each bird finds its own place to spend the night: a hole in a tree trunk, a gap under a loose piece of bark, or a small space between the needles of a pine tree.

It's near dark, and this black-capped chickadee has found a tree where it can go dormant for the night.

Once a chickadee nestles into its cubbyhole, it fluffs up its feathers and tucks its beak under its wing. Then it goes into hypothermia. Its temperature drops from 108°F (42.2°C) to about 88°F (31°C). Its heart rate and breathing slow way down.

IT'S A FACT!
A chickadee would have to eat all night to keep its temperature as high as it is during the day. By going dormant, the bird saves energy and gets a good rest.

The chickadee may be dormant, but it is still in control. If the bird gets a bit too cold, it shivers. That raises its temperature a few degrees. In the morning, it revs up its heart and raises its temperature. When it is all warmed up, the chickadee comes out of its hole and rejoins the flock.

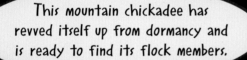

This mountain chickadee has revved itself up from dormancy and is ready to find its flock members.

Human Hypothermia

You can go into hypothermia, too, but you can't control it as chickadees can. If your body temperature drops to 95°F (35°C), you start to shiver and stamp your feet. If that doesn't warm you up, you keep getting colder. At around 90°F (32°C), your heart slows down. Your muscles get stiff, and you act clumsy and confused. If your temperature drops another few degrees, you pass out. Your heart sputters to a stop. If you don't get help soon, you will die. But if you do get help to warm up, you can survive hypothermia.

IT'S A FACT!

Even when you feel very cold, your body temperature is close to its normal level of 98.6°F (37°C). It drops a couple of degrees while you're sleeping, but it doesn't change at different times of the year.

Government scientists set up drilling equipment in Antarctica. Even though they are wearing warm clothing, the scientists can tolerate the cold temperatures for only so long before deadly hypothermia sets in.

Miracles in the News

Hypothermia can be deadly, but it has saved people who have fallen into cold water. You might have heard a story like this on the news. A car skids off an icy road and crashes into a pond. A little

girl riding in the backseat gets thrown into the water. When rescuers pull her out half an hour later, she looks dead. She's cold all over, her heart isn't beating, and her lungs are filled with water. But the rescue team does CPR and gradually warms her up. Suddenly, she takes a breath and then another. She's alive!

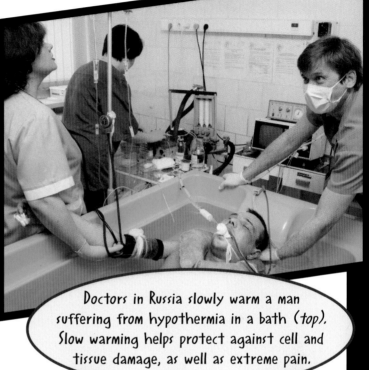

Doctors in Russia slowly warm a man suffering from hypothermia in a bath (*top*). Slow warming helps protect against cell and tissue damage, as well as extreme pain.

IT'S A FACT!

People who suffer severe hypothermia in air, rather than in water, usually don't survive (*right*). They cool too slowly. Their brain cells continue to need oxygen but can't get it because their heart stops working.

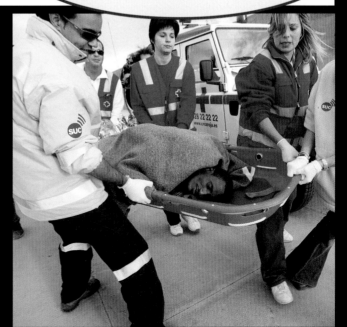

Why didn't this girl die? She lived because she went dormant. Doctors think that when she breathed in the cold water, it cooled the blood passing through her lungs. Then the blood cooled her brain. That made her brain cells go dormant, so they needed less oxygen. She survived because her brain got very cold very fast.

Putting It to Use

Doctors learned a lot from cases like that. They learned to put patients into hypothermia on purpose. For surgery or other medical procedures, doctors usually make patients sleep with drugs called anesthetics. With hypothermia, doctors can keep patients asleep with less anesthetic. For some patients, lying in an ice bath is enough. Their body temperature drops just enough to slow down the brain cells a bit.

Other patients must go deeper into dormancy before surgery. Doctors drain out their blood and replace it with a chilled fluid. The heart stops, and the brain goes quiet. Since the heart isn't moving, it's easier to operate on. And since the brain is dormant, it's not hurt by the lack of

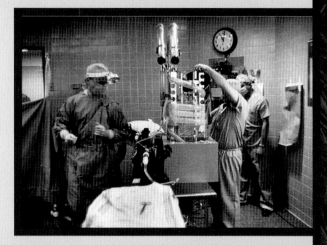

Doctors, Start Your Engines
Since the late 1950s, doctors have used a special heart-lung machine (right center) to remove and cool a patient's blood. Before that machine was invented, one doctor tried using a car radiator to cool the patient's blood.

Hypothermia and Healing

Using hypothermia during surgery is nifty, but it can be dangerous too. A patient who cools too slowly or too quickly could die or suffer brain damage. Even if everything goes well, patients who are put into hypothermia lose more blood, heal more slowly, and get more infections than patients who stay warm during their operations.

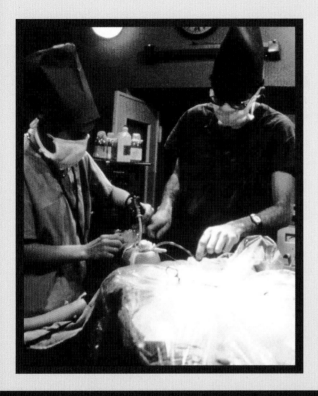

These doctors are using hypothermia during heart surgery. Deep hypothermia completely stops the flow of blood in the patient temporarily.

blood and oxygen. When the operation is done, doctors funnel the blood back into the body and slowly warm up the patient.

Taking the blood out of a person's body is awfully drastic. It also requires big, heavy equipment. Doctors are working on a way to cool a patient with small, portable equipment that emergency rescue teams can carry. This new technique copies what happens when a person drowns in cold water. A machine pumps an icy-cold fluid into the patient's lungs. There, the fluid cools the patient's blood. It then moves on to cool the brain and the rest of the body.

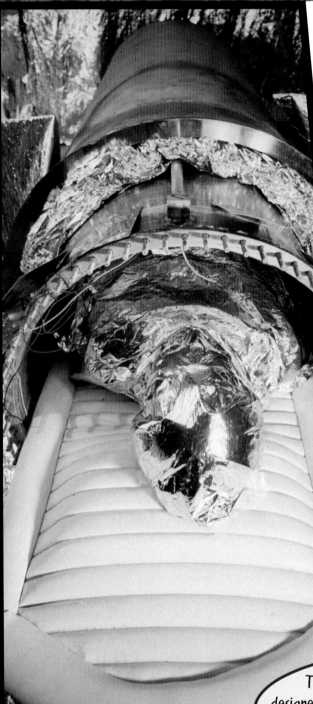

This technique could be a big breakthrough for heart attack and stroke victims. Often the worst damage they suffer doesn't happen during the attack but soon afterward. If emergency teams could make the patient go dormant until they reach a hospital, the patient would have a much better chance of surviving.

Hypothermia is a great tool for doctors and emergency teams. But it's short-term. Patients can stay cold and dormant for just a few hours. What about those who want to stay dormant for months or even years?

This is a model of the cryochamber designed for a cancer patient in 1967. The man remains in cryogenic storage, awaiting revival and a cure for cancer.

chapter 5

Nap Time

Below the chickadee's tree and under the wood frog's leaf pile, the groundhog settles into its den. It will stay there all winter. Like many mammals, the frog hibernates when the weather turns cold. If it lives in a place with very long winters, the frog might spend more than half of every year in hibernation.

Its temperature, which is about 99°F (37°C) when it's awake, drops to about 54°F (12°C). Its heart beats just a few times a minute. The frog takes a breath once every few minutes, but sometimes it doesn't breathe for more than an hour.

The groundhog pigs out before entering hibernation. It puts on enough fat to fuel its cells during the long winter. While hibernating, it uses very little energy. When warm weather returns in spring, the groundhog takes a few days to wake up all the way. When it does, it's hungry but healthy.

Sleeping Like a Bear

Different animals hibernate in different ways. Arctic ground squirrels get very cold, and they sleep so deeply that they're actually brain-dead. Bears only cool down a little bit, and their brains stay active. Their hibernation is more like regular sleep. That might be how we would hibernate, if we could.

Denning bears, such as this black bear, are shallow hibernators. Though they sleep deeply and their heart slows, their body temperature stays nearly normal (88 to 95°F, or 31 to 35°C).

Wake Me When We Get There

Hibernation looks like the perfect way for us to go dormant while we wait for a medical breakthrough or take off on a voyage to other planets. We wouldn't have to worry about ice damage or toxic cryoprotectants. We'd just go to sleep but for longer than usual, right?

Maybe. But we have a lot to learn first.

For starters, we don't know what makes hibernating animals hibernate. It's not just a matter of getting colder. That would be hypothermia, where your temperature drops, and that makes your metabolism slow down. When an animal goes into hibernation, its metabolism drops within half an hour, but its temperature keeps falling for several more hours.

IT'S A FACT!

Hibernation is not caused by a cold environment or not having food. Many animals get cold or hungry and never enter hibernation. There's something special about hibernators.

Some scientists think the metabolism slows down because of a substance in the blood called a hibernation trigger. If there is such a substance and if we can find it, maybe we could use it to put ourselves into hibernation. Scientists have looked for a trigger, but so far, they haven't been able to find one. Whatever makes hibernating animals hibernate remains a mystery.

We also don't know what makes hibernating animals wake up. We definitely need to know that before we try to make people hibernate!

Off on a Journey

You might have seen movies in which astronauts hibernate during long space voyages. Let's take a closer look at what that would be like. Say you've been chosen for a mission to Mars, and NASA plans to put you into hibernation for the 18-month flight.

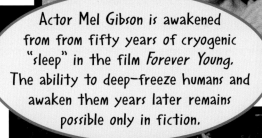

Actor Mel Gibson is awakened from from fifty years of cryogenic "sleep" in the film *Forever Young*. The ability to deep-freeze humans and awaken them years later remains possible only in fiction.

First, you'll need to gain weight. If you normally weigh 100 pounds (45 kilograms), you'll have to add 100 pounds of pure fat. That's what you'll live off while you're dormant. Without it, you will starve.

But all that fat will be gone after a few months—and you'll have about a year of hibernation yet to go. What then? Maybe you could be fed through tubes, like some hospital patients. No one will be around to refill your supply line or take care of problems, so the system needs to work perfectly.

Early cryobiologists demonstrate the freezing process for human cryostorage in 1967. If this were an actual, successful cryopreservation, the patient would likely wake to a very unfamiliar world.

DADLE

Doctors have developed a drug called DADLE that is similar to a hibernation trigger. It slows the metabolism. With DADLE, doctors can keep animal hearts healthy outside the body for nearly two days. Without DADLE, the hearts only last about 14 hours. If DADLE does the same with human hearts, it could help us store them longer before they are transplanted.

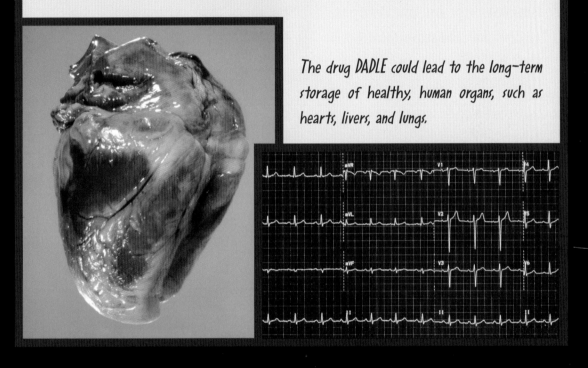

The drug DADLE could lead to the long-term storage of healthy, human organs, such as hearts, livers, and lungs.

During the flight, you'll roll over and change positions every few weeks. But you won't wake up enough to steer the ship, run experiments, or send messages to your family.

If anything goes wrong, you won't be able to fix it. Even if Mission Control gives you an emergency shot of "wake-up trigger," you won't be able to do much right away. When you come out of hibernation, you'll be groggy and uncoordinated for a couple of days. Your muscles and bones will probably be weak. You might have trouble walking.

When it's time to come home, you'll have to pack on the pounds again and hook yourself up to feeding tubes before settling in for the long ride home.

Hibernating on spaceflights might be possible someday. But we have lots of practical problems to solve before we can make the dream of human hibernation come true.

Astronauts sleep on the space shuttle. Extended spaceflights and human hibernation are still just dreams. But advances in cryobiology suggest the dreams could become reality.

chapter 6

Into the Future

When you read about all the problems with trying to go dormant, suspended animation seems like an impossible dream. We can't revive people after freezing them for one minute, much less 100 years. We can't even make a human heart go dormant and stay alive for more than a few hours. We're still in the dark about what makes animals hibernate.

But we have made progress. We understand a lot about how cryoprotectants and ice blockers work. We can freeze and revive the embryos of mice, people, and endangered species. We can put human patients into hypothermia, and we've made a drug that acts a lot like a hibernation trigger.

USE RESTRICTED
LIVE
HUMAN
EMBRYOS!

A technician places live human embryos into liquid nitrogen cryostorage.

Just Add Water

We've talked about going dormant by getting colder. Another way of going dormant doesn't depend on the temperature. Some plants, insects, and small creatures do it by drying out. They shrivel up into crusty little nuggets that can survive for more than 100 years. Just add water and—foom! They swell up and come back to life. This approach isn't very promising for use with people. It only seems to work with tiny animals. Human cells usually die if they dry out. Besides, we probably couldn't get people to try it. Being frozen is gross enough. Can you imagine being turned into a giant jerky stick?

Flower seeds bloomed after being dormant for as many as 100 years after heavy rains fell on the dessert in Death Valley, California, in 2005 (above). Similarly, a dried worm (inset) has the ability to stay dormant for many years, coming to life with water.

We are closer to discovering the secrets of going dormant than we ever have been before. Suspended animation may be a dream, but it is becoming a possible dream.

Waking Up Is Hard to Do

When we solve the scientific questions about how to go dormant, we will face a different kind of question, a more personal one. Would you want to be put into suspended animation? Imagine what it would be like. While you're dormant, you won't see, hear, think, or dream. It won't be like being alive at all.

What about waking up? Who decides when to bring you back? What happens then? Imagine you've been dormant for 100 years. How will you live? What kind of job will you be able to get?

The technology will be completely different than when you went dormant. You won't have a clue how anything works. Think of people from 100 years ago trying to program a computer or run a DVD player. If you're dormant for 100 years, when you wake up, all you'll know is that everything has changed, except you.

Worst of all, your friends and family will be gone. Who will celebrate your new life with you?

Helping People Today

Freezing people for a long time gets headlines, but other uses of suspended animation could help many more people. Look at organ transplantation. Every day in the United States, about 70 people receive an organ transplant that saves their life—and 16 people die because they didn't get the organ they needed. Dozens of organs die too. If an organ

Organ Banks

Modern organ banks help bring donated tissues and organs to patients who need them. Organ banks freeze or vitrify small tissues and organs, such as blood vessels (right), and store them for months or years. But they can't store organs such as hearts and livers for more than a few hours. We don't yet know how to put these organs into suspended animation and bring them safely back to life.

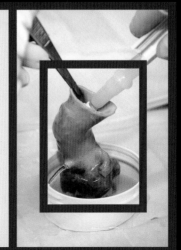

A surgeon places a living human kidney into a bag of ice. It is then packed with additional ice in a box. Because the ice won't make the kidney go dormant but only slow its death, the kidney must be rushed to a needy recipient within hours.

can't be matched with a patient in time, it goes to waste.

The problem is that we can't make the organs go dormant. We can only keep a human heart or lung healthy for about 6 hours. A liver lasts 24 hours, and a kidney lasts 72 hours. Often we don't have time to match an organ with a patient. Or we know they match, but the organ is too far away from the patient. By the time we moved it there, the organ would be dead.

If we could make organs go dormant for a week, we could save thousands of lives every year. Just think what we could do if we could store organs for months or years. We'd have organ banks like The Jackson Lab's mouse bank. People who needed a transplant could get one from the bank. They wouldn't have to wait and hope that a matching organ would become available in time to save them.

On the Frontier

As we learn more about going dormant, we will invent new uses for this amazing skill—uses we currently can't predict. Who would have thought that doctors could save patients by chilling them to stop their heart and brain? Who would have dreamed of preserving organs by turning them into glass? The coolest thing about research isn't what we can do today. It's what we might come up with tomorrow.

That's where you come in. You might be the person who discovers what makes bears hibernate. Maybe you'll invent a new antifreeze that lets us store hearts for a year.

And while you're in the lab cooking up a new cryoprotectant or hibernation trigger, remember where the work started—with the chickadees, ladybugs, and other animals that are the real masters of suspended animation.

Continued research in cryobiology may lead to the ability to store complex human organs, such as the brain. It may even result in our ability to successfully cryopreserve the human body for deep space travel or life in the future.

Glossary

anesthetic: a drug that makes a patient sleep deeply enough to be operated on

antifreeze: a chemical that prevents freezing

chill damage: injury that happens to cells and organs whose temperature falls several degrees below their normal temperature. Chill damage can be fatal.

cryonics: the field of research that deals with cold preservation of human bodies, with the goal of reviving them in the future

cryopreservation: low-temperature storage of organs, tissues, or cells

cryoprotectant: a substance that protects cells and organs from being damaged by ice crystals

diabetes: a disease in which the blood sugar called glucose rises to dangerous levels

dormant: not active

freezing point: the temperature at which a liquid changes into a solid

frostbite: injury caused by freezing, usually of fingers, toes, ears, or nose

glucose: a sugar that is carried in the blood and used by cells for energy. Some animals use glucose as a cryoprotectant.

hibernation: a sleeplike state used by many mammals and some birds to endure lengthy periods without food

hypothermia: a lower-than-normal body temperature

metabolism: all the chemical processes that go on within a living organism

nucleator: a particle that speeds the formation of ice crystals. Dust, tiny crumbs of food, and bacteria can all act as nucleators.

supercool: to lower the temperature of an animal or a liquid below the usual freezing point without the formation of ice

suspended animation: the temporary slowing or stopping of vital body functions, such as breathing and heartbeat

vitrification: the process of changing a liquid into a hard substance that does not contain crystals

Selected Bibliography

Asymptote. *The Asymptote Cool Guide to Cryopreservation.* N.d. http://www.asymptote.co.uk/process/cryo/cryoguide/report/guidehomepage.htm (March 2005).

BioPreservation, Inc. *Cryopreservation.* N.d. http://www.cryocare.org/index.cgi?subdir=bpi&url=bpi.html (March 2005).

Boyer, Bert B., and Brian M. Barnes. "Molecular and Metabolic Aspects of Mammalian Hibernation." *BioScience* 49 (1999):713–724.

"Cryobiology." *Beauchamp Search.* N.d. http://www.beauchamp.de/odp/odp.php/browse/ Science/Biology/Cryobiology (March 2005).

Fahy, Gregory M., et al. "Cryopreservation of Organs by Vitrification: Perspectives and Recent Advances." *Cryobiology* 48 (2004):57–178.

Lee, Richard E., and Jon P. Costanzo. "Biological Ice Nucleation and Ice Distribution in Cold-Hardy Ectothermic Animals." *Annual Review of Physiology* 60 (1998):55–72.

New England Organ Bank. "Organ and Tissue Transplants Work." *Transplantation.* N.d. http://www.neob.org/trans.html (March 2005).

Oeltgen, Peter R., et al. "Extended Lung Preservation with the Use of Hibernation Trigger Factors." *Annals of Thoracic Surgery* 61 (1996):1,488–1,493.

Storey, Kenneth B., and Janet M. Storey. "Natural Freeze Tolerance in Ectothermic Vertebrates." *Annual Review of Physiology* 54 (1992):619–637.

Fullick, Ann. *Rebuilding the Body: Organ Transplantation.* Chicago: Heinemann Library, 2002. With this book, you can learn how doctors decide an organ transplant is needed, how donor and recipient are matched, how donated organs are preserved, and more.

Heinrich, Bernd. *Winter World: The Ingenuity of Animal Survival.* New York: Ecco Press, 2003. Follow many species of wildlife, from insects to chickadees, as they face the trials of winter.

Nirgiotis, Nicholas, and Theodore Nirgiotis. *No More Dodos.* Minneapolis: Lerner Publications Company, 1996. Learn how zoos are using frozen sperm and embryos to preserve endangered species.

Wharton, David A. *Life at the Limits.* Cambridge, UK: Cambridge University Press, 2002. Chapter 5, "Cold Lazarus," is loaded with cool information about how animals cope with the cold.

The Coalition on Donation

http://www.shareyourlife.org

Get answers for your questions about organ donation at this website from the Coalition on Donation. The site includes videos and a Web documentary.

Hypothermia.org

http://www.hypothermia.org/

Learn about hypothermia—its warning signs, how to avoid it, and what to do if you or someone you're with becomes hypothermic.

The Jackson Laboratory

http://www.jax.org

Take a virtual tour of The Jackson Laboratory. See how scientists there freeze (and thaw) mouse sperm and embryos in the lab's "mice bank."

Laboratory for Ecophysiological Cryobiology at Miami University

http://www.units.muohio.edu/cryolab/

Find out about research projects on how frogs, snakes, turtles, and insects make it through winter.

Index

Photo Acknowledgments

Photographs are used with the permission of: PhotoDisc Royalty Free by Getty Images, pp. 1, 2–48 even pages (background); © Matt Meadows/Peter Arnold, Inc., p. 4; © 20th Century Fox/The Kobal Collection/The Picture Desk, p. 5; © Dr. John Brackenbury/Photo Researchers, Inc., p. 6; © National Geographic/Getty Images, p. 7; © Merlin D. Tuttle, Bat Conservation International, p. 8 (left); © Asgeir Helgestad/naturepl.com, pp. 8 (right); © Gerald and Cynthia Merker/Visuals Unlimited, p. 9 (left); © age fotostock/SuperStock, p. 9 (right); © Jens Schlueter/AFP/Getty Images, p. 10 (top); © Martin Hartley/eyevine/ZUMA Press, p. 10 (middle); © Dr. Gladden Willis/Visuals Unlimited, p. 10 (bottom); © Michael P. Gadomski/SuperStock, p. 13; © Gustav Verderber/Visuals Unlimited, p. 14 (left); © SIU/Visuals Unlimited, pp. 14 (right), 15 (right), 37 (left); © Biodisc/Visuals Unlimited, p. 15 (left); © Klaus Guldbrandsen/Photo Researchers, Inc., p. 16 (left); © Mauro Fermariello/Photo Researchers, Inc., p. 16 (center and right); © Dr. Dennis Kunkel/Visuals Unlimited, p. 17 (inset); © Ralph Hutchings/Visuals Unlimited, p. 17 (main); Bill Hauser, p. 18; courtesy 21st Century Medicine, p. 19 (left), 24 (all), 25; © Suzette Van Bylevelt/ZUMA Press, p. 19 (right); photo courtesy of Alcor Life Extension Foundation, p. 20 (left); © Michael Macor/San Francisco Chronicle/CORBIS, p. 20 (center and right); © Dr. Ken Wagner/Visuals Unlimited, p. 21; © Mike Anich/Visuals Unlimited, p. 22; © Charles George/Visuals Unlimited, p. 23; © Marcia Griffen/Animals Animals-Earth Scenes, p. 26; © D. Robert and Lorri Franz/CORBIS, p. 27; Melanie Conner/Antarctic Photo Library, National Science Foundation, p. 28; © Oleg Nikishin/Getty Images, p. 29 (top); © Samuel Aranda/AFP/Getty Images, p. 29 (bottom); © Al Fenn/Time Life Pictures/Getty Images, p. 30; © NOVOSTI/Photo Researchers, Inc, p. 31; © Henry Groskinsky/Time Life Pictures/Getty Images, p. 32; © Stephen Lang/Visuals Unlimited, p. 34; © Warner Bros./The Kobal Collection/The Picture Desk, p. 35 (both); © J.R. Eyerman/Time Life Pictures/Getty Images, p. 36; © Dr. P. Marazzi/Photo Researchers, Inc., p. 37 (right); courtesy of NASA, p. 38 (both); © Jim Olive/Peter Arnold, Inc., p. 39; © David McNew/Getty Images, p. 40 (main); © Vaughan Fleming/Photo Researchers, Inc., p. 40 (inset); © Antonia Reeve/Photo Researchers, Inc., pp. 41, 42 (right); © Will and Deni McIntyre/Photo Researchers, Inc., p. 42 (left); © Alexander Tsiaras/Photo Researchers, Inc., p. 43.

Front cover (clockwise from top): courtesy of Carlisle P. Landel, Ph.D./The Jackson Laboratory; © Alexander Tsiaras/Photo Researchers, Inc.; Evelyn Davidson, courtesy of the Laboratory for Ecophysiological Cryobiology, Miami University, Oxford, Ohio; © Charles George/Visuals Unlimited; PhotoDisc Royalty Free by Getty Images (background). Back Cover: PhotoDisc Royalty Free by Getty Images (background)

About the Author

Cherie Winner holds a Ph.D. in zoology from Ohio State University. She is a full-time science writer for *Washington State Magazine.* Winner also writes books. Her published titles include *Salamanders, Coyotes, Trout, Woodpeckers, Erosion,* and *Life in the Tundra.*